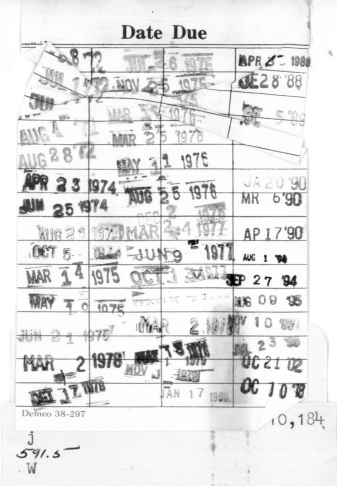

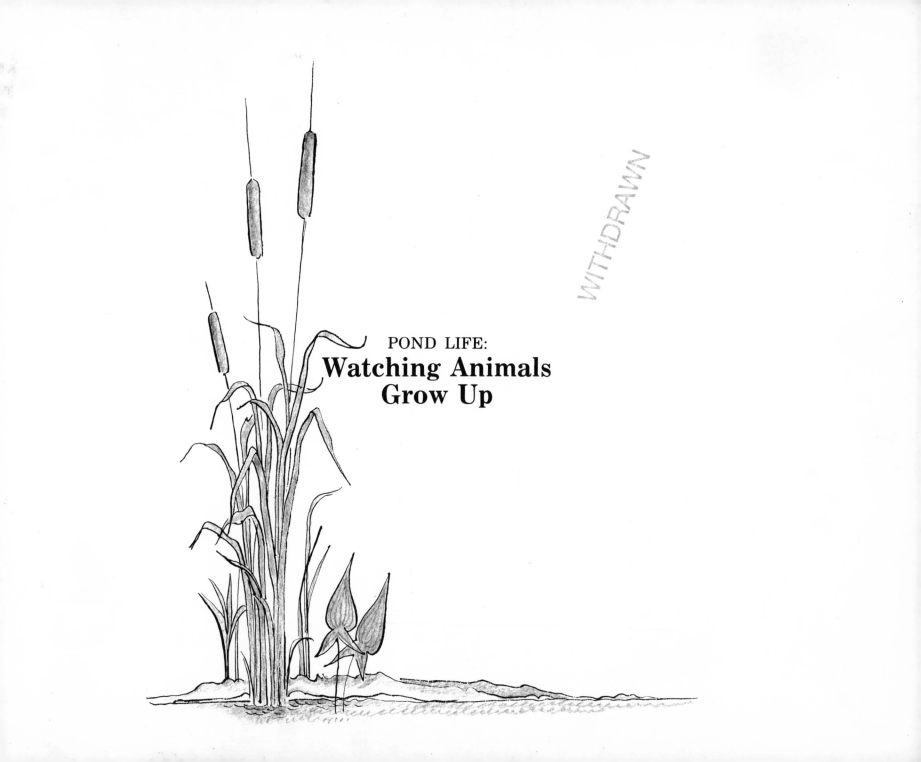

POND LIFE:
Watching Animals Grow Up

POND LIFE:

Watching Animals Grow Up

by Herbert H. Wong
and Matthew F. Vessel

illustrated by
Harold Berson

▲ Addison-Wesley

Science Series for the Young

Level A
My Ladybug
My Goldfish
Our Tree
Our Terrariums

Level B
Pond Life: Watching Animals Find Food
Animal Habitats: Where Can Red-Winged Blackbirds Live?
Plant Communities: Where Can Cattails Grow?
Pond Life: Watching Animals Grow Up

An Addisonian Press Book

Text copyright © 1970, by Herbert H. Wong and Matthew F. Vessel
Text Philippines copyright 1970, by Herbert H. Wong and Matthew F. Vessel
Illustrations copyright © 1970, by Harold Berson
Illustrations Philippines copyright 1970, by Harold Berson

The Addison-Wesley Publishing Company, Inc.
Reading, Massachusetts
Library of Congress catalog card number 72-118993
Printed in the United States of America
First Printing
SBN: 201-08730-8

SCIENCE
SERIES
FOR THE
YOUNG

I can hear red-winged blackbirds.
We must be by the pond.

There is a turtle! And turtle eggs!
The mother turtle puts sand over the eggs.

Will the mother turtle take care of them?
No. She will not stay with them.
There she goes now.
She is going into the pond.

9

The turtle swims by some sunfish.
Is she going to catch one of the baby sunfish?

No, not this time.
A big sunfish swims over.
It keeps the turtle away from the baby sunfish.
The turtle swims on.

Do all the fish take care of baby fish like that?
Some do, but not many.

Look at that dragonfly go — up and down!
What a funny way to fly!
Dip . . . dip . . . dip.
She puts her eggs into the water that way.
At every dip, another egg goes into the water.

But where are the dragonfly babies?
Here are some.
These babies are about four days old.
They can swim and find food.
But they are not ready to fly.

This dragonfly is about ready to fly.
It has come out of the water.
Soon it will get wings.
A dragonfly changes when it grows up.

Do other animal babies change like that?
Some do. Some do not.
See those little water skaters?
They look just like big skaters.
They do not have to change.
Water skaters just grow.

15

But mosquitoes make a big change.
Do you see the little things in the water there?
They are eggs.
Next to them are baby mosquitoes, just hatched.
They do not look like mosquitoes at all.
After about three days they change.

In another week, they will change again.
They will get wings and fly away.

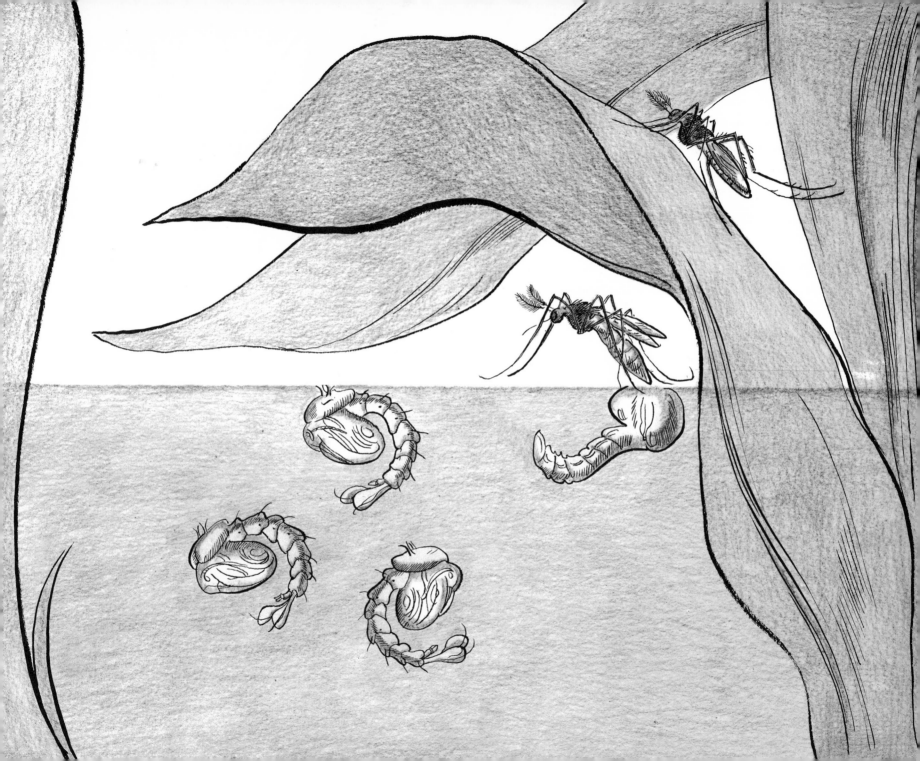

Where did the turtle go?
I cannot see her.
But I see a bird nest over there.
Can you see any eggs?
No, the nest is too far away.
But red-winged blackbirds are there.
There must be eggs in the nest.
Red-winged blackbirds stay with the eggs.
They keep the eggs warm.

When will the eggs hatch?
Not for some time.
The baby birds come out in about two weeks.
But there may be other baby animals
around the pond now.
I see some ducks!
Are some of them baby ducks?

Yes! The baby ducks are not very big.
But they can swim and find food.
The mother duck stays with them.
She takes care of them and keeps them warm at night.

Hey, I found some snail eggs!
You can see baby snails in some of them.
And some of the baby snails are coming out!
Where are the big snails?
Some are over there.
But they do not take care of the baby snails.

What was that?
It ran under there.
A mouse! And babies!
The mother mouse gives them food.
She keeps them warm, too.

Hey, there is the turtle, on that old tree.
What is that under the tree?
A big ball of frog eggs!

Why don't we take some frog eggs home?
Then we can see them hatch.

Mother, look what was in the pond!
Frog eggs!
We want to see the frogs grow up.

After four days tadpoles come out of the eggs.
They do not look like frogs.

But then they change. They get legs.
The back legs come first.
The tadpoles look funny.
Now the tadpoles have four legs.

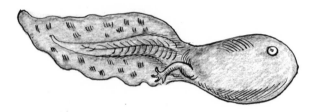

Now they are grown up frogs.
It is time to take them back to the pond.

We saw many babies the other time we came to the pond.
Are they here now?

This is where the mouse nest was.
Are the babies there? No, they are not.
Baby mice grow up fast.
They can find food with no help now.

There are the baby ducks.
They are not so little now.
They look like the big ducks.

What other babies were there?
The red-winged blackbirds!
Have the eggs hatched?

Where is the nest? I see it!
And I can see baby birds!

There is the mother bird, with some food.
The baby red-winged blackbirds will be ready
to fly soon.

Hey, where are the baby turtles?
This is where the eggs were.
They were under the sand.
Maybe they are here now.
Maybe they have not hatched yet.
We can come back again some day and see.

Boy, is it warm today!
We have not come to the pond in a long time.
The pond babies must be grown up now.
There is the red-winged blackbird nest.
There are no babies in it.
But there are many red-winged blackbirds.
Were some of them the babies we saw?
Yes, I think so.

Where are the ducks?
There they are.
The baby ducks are grown up now, too.

Hey, look!
There are the baby turtles.
Here they come, out of the sand.
They look just like grown turtles.
And they can swim!

There are many animals at the pond.
And we saw some of them grow up.
Some take a long time. Others grow up fast.
Some babies do not have any care at all.
Grown animals take care of others.

Some animals make big changes when they grow up.
Some make little changes.
And some babies just grow up.

We came to the pond many times.
We saw many things.
And we will come again.

About the Authors

Herbert H. Wong is well known as an author of science books for children and teachers, as a science education consultant, and as an educator. He is the Principal of the University of California Laboratory School, Washington Elementary, in Berkeley, California. Dr. Wong is active in many current science curriculum development projects and pilot programs. He holds a degree in Zoology and Ed.D. from the University of California.

Matthew F. Vessel, Associate Dean of the School of Natural Science and Mathematics of San Jose State College, is an author, editor, and consultant, in science and education. He is a Fellow of the American Association for the Advancement of Science. An active member of many other professional science and education societies, Dr. Vessel attended St. Cloud Teachers College and the University of Minnesota. He holds the Ph.D. from Cornell University.

About the Artist

Harold Berson derives great enjoyment from drawing people and landscapes and traveling to find just the right combination. Both he and his wife like North Africa inordinately and have enjoyed extended stays in Tunisia and Morocco. The Bersons have also visited Turkey, Yugoslavia, and France, sketching and watercoloring wherever they went. Both Bersons are very interested in animals and nature and often visit zoos and museums to sketch animals. They live in New York and love it, in spite of it all.